Mick Manning &
Brita Granström

KINGfISHER

KINGFISHER
Kingfisher Publications Plc
New Penderel House
283-288 High Holborn
London WC1V 7HZ

First published by Kingfisher Publications Plc 1998
This edition published 1999
10 9 8 7 6 5 4 3 2
2TR(1BFC)/0399/WKT/P&W(P&W)/140KTMA

A CIP catalogue record for this book is available from the British Library.

ISBN 0 7534 0210 6 (hb)
ISBN 0 7534 0252 1 (pb)

Printed in Hong Kong/China

For Ivar Andersson
1912-1996
Love Brita and Mick

Series designer: Terry Woodley
Editor: Max Benato
Designer: Nina Tara
Text: Sally Morgan
Consultant: Mukul Patel

Contents

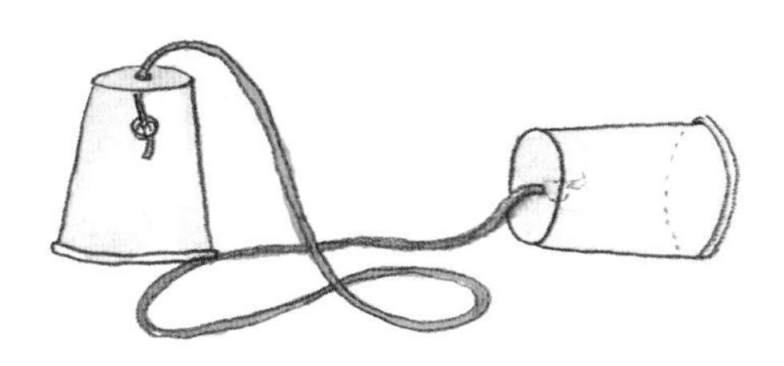

What this book is about

Do you know how to bend light or why puddles quickly disappear? **Science School** is full of exciting experiments that will help you to answer these questions and many more. **Science School** teaches you how to have fun with science!

Some of the things you will need:

SCIENCE TIPS

Helpful tips appear in the coloured strips on each page.

Look in the glossary at the back of the book to find out about words in ***bold type****.*

Being a scientist

Scientists learn about the world we live in by carrying out experiments. Sometimes these experiments lead to exciting new discoveries and inventions that change the world. Scientists invented the wheel, told us the Earth is round, and gave us computers and space rockets!

Eureka!

WOW!

Keeping a NOTEBOOK

PREPARATION

Scientists are very organized and careful people. When they carry out an experiment, they write down what they used, what they did and what happened. You should keep a notebook throughout *Science School* so that you can write about the experiments you do step-by-step – you can even draw tables and pictures to show your results. Have fun and don't worry if your experiments don't always come out as you would expect – some of the greatest scientific discoveries were made by mistake! If an experiment doesn't work, try and figure out what went wrong and write about it in your notebook.

SCIENCE TIPS

Why not look these famous scientists up in the library or on the Internet?

Sir Isaac Newton figured out that the same force that makes an apple fall from a tree makes the planets move round the Sun.

Leonardo da Vinci drew a helicopter centuries before it was invented!

Marie Curie won two Nobel Prizes for her discoveries in the world of physics.

SAFETY TIP

Scientists know how important it is to be careful when carrying out experiments. Never taste any chemicals and always ask an adult to help you with the trickier experiments. Never use sharp scissors or hot liquids without an adult's help and be extra careful with heat and electricity.

Matter

Everything around us is either a solid, a liquid or a gas. Solids such as metals are usually hard and have a set **volume** (size) and a set shape. Liquids such as water have a set volume but take the shape of whatever container they are in. Gases such as air have no set volume or shape. Scientists call solids, liquids and gases the three states of **matter**.

SCIENCE TIPS

The steam rising from a kettle of boiling water is water vapour, or gas, escaping from the liquid.

Dry ice is used in films and on stage to create eerie and mysterious scenes. It is made by heating solid carbon dioxide. Instead of melting into a liquid, this changes straight from a solid into a gas.

Making clouds

Put some crushed ice in a large tin with about one-third as much salt. Push a smaller tin into the ice mix, being careful not to touch the ice with your fingers. Blow into the small tin. You will see a small cloud. This is because your breath contains a lot of water **vapour** (gas). When this vapour hits the cold air in the small tin, it turns from a gas into tiny drops of water, which make up the cloud.

Magic ice

How do you get an ice cube out of a glass of water – without getting wet? Dangle the end of a piece of string on to the ice cube. Sprinkle salt on the ice and leave it for a few minutes. Pull the string to lift the ice out of the glass. The salt makes the ice melt a little and the string gets soaked in the water. But the water soon freezes again, trapping the string.

Evaporating perfume

Place a tiny drop of perfume on your arm. Perfume contains alcohol which evaporates on warm skin – taking heat away and cooling the skin.

SCIENCE TIPS

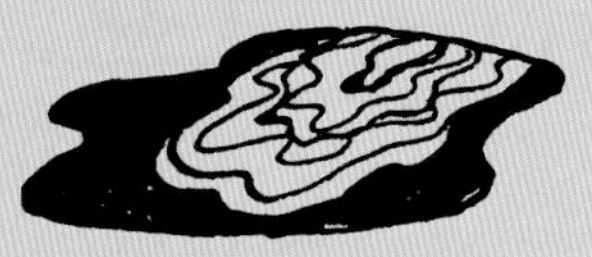

On a hot day, the water in a puddle will quickly evaporate, or turn into tiny drops of water vapour.

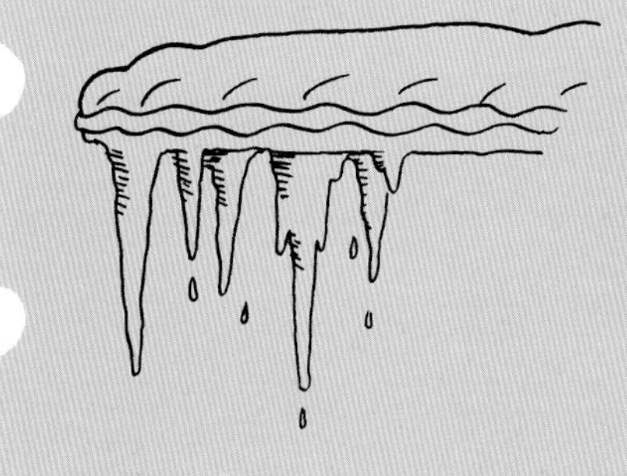

On a very cold night, water dripping from a roof will freeze and form icicles.

Chocolate leaves

1.

2.

3.

EXPERIMENT

You will need: ● chocolate ● a small bowl ● a saucepan ● a heatproof glove ● a clean paintbrush

1. Pick some rose leaves, wash them under the tap and dry them carefully.

2. Ask an adult to pour some hot water into a saucepan and put the bowl inside, making sure the water does not get into the bowl. Put some pieces of chocolate in the bowl. The heat from the water will melt the chocolate, turning it into a thick liquid. Ask an adult to remove the bowl of chocolate with a heatproof glove.

3. Use the paintbrush to paint the chocolate on the top side of each leaf.

4. As the chocolate cools, it turns back into a solid and takes up the shape of the leaf. Carefully peel away the real leaves.

Getting a reaction!

Fireworks burning in the air, a cake baking in the oven and a car going rusty in the rain – these are all examples of chemical reactions. Chemical reactions happen when one substance combines with another, making an entirely new substance. They happen all around us and even inside us – the millions of reactions that take place in our body keep us alive.

Stand back and wear old clothes!

Make a volcano...

EXPERIMENT

See what happens when a substance called an **acid** meets another substance called an **alkali**.
1. Put some bicarbonate of soda in a small bottle and bury it in some sand shaped like a volcano. You should be able to see the top of the bottle. **2.** Mix some tomato ketchup with vinegar in a bowl. **3.** Pour the mixture into the bottle. **4.** Stand back and watch the substances react! The chemical reaction produces a gas which forces the mixture out of the bottle – like a volcanic eruption!

1.

4.

2.

3.

Shiny money

EXPERIMENT

Copper coins often look dull and dirty because oxygen in the air reacts with the copper, making a copper **oxide** coating. Take a dirty copper coin and soak it in some lemon juice for a few minutes. What happens? The acid in the lemon juice reacts with the oxide and removes it – leaving a shiny copper coin!

SCIENCE TIPS

Iron nails go rusty when they are left in the rain because the iron reacts with the water and oxygen in the air.

Bicycle frames that are made out of iron often have a protective layer of paint to stop them going rusty.

Acids and alkalis!

EXPERIMENT

Many chemical reactions involve **acids** and **alkalis**. You can tell if something is an acid, like vinegar, or an alkali, like bicarbonate of soda, by testing it with red cabbage. Ask an adult to help you.
You will need: • a red cabbage • water • a saucepan • a strainer • a lemon • soap
1. Ask an adult to cook some cabbage in water. **2.** When the cabbage has cooled, strain the purple juice into a jug. **3.** Pour a little juice into a glass and squeeze some lemon juice into it. **4.** Pour some juice into another glass that has a piece of soap in it. If the colour of the juice turns pink, the substance is an acid. If the colour of the juice turns greeny blue, it means the substance is an alkali.

3.

4.

SCIENCE TIPS

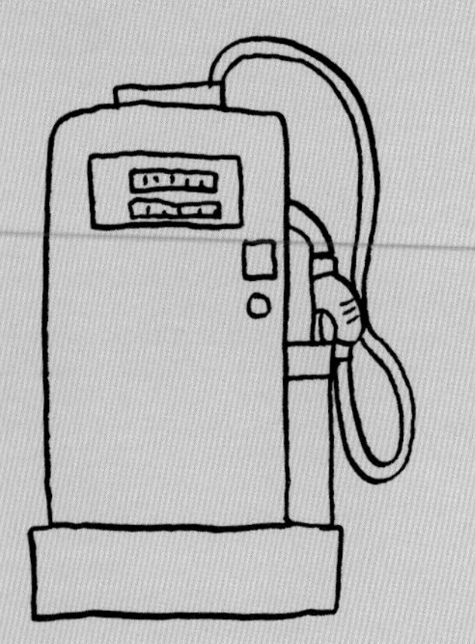

Petrol comes from crude oil. It is separated from the oil by a process called refining. Diesel oil and bitumen, used to make roads, can also be separated from crude oil.

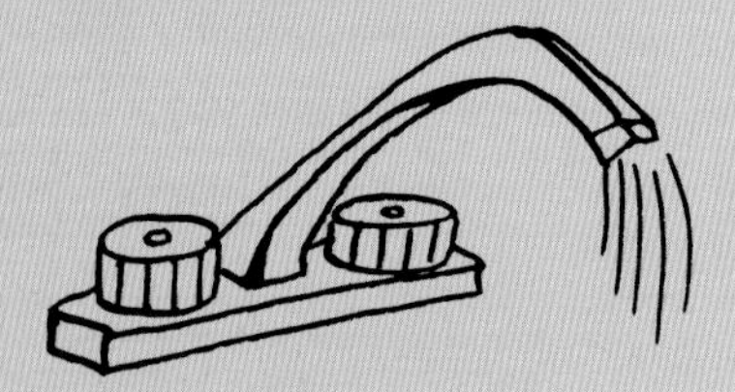

Before it reaches our homes, tap water is thoroughly filtered so that any dirt is separated and removed.

Cars have both oil and air filters to stop dirt getting into the engine and damaging it. Eventually, the filters get clogged up with dirt and have to be changed.

Separating things

Many things are made from several different substances which can usually be separated quite easily. These are called **mixtures**. Milk, for example, is a mixture of water, fat and minerals such as calcium. There are lots of different ways to separate the substances in a mixture.

Oil and vinegar

EXPERIMENT

Salad dressing is often made from just oil and vinegar but when oil and vinegar are shaken together they separate into layers. Take a small bottle and pour in some oil. Now add the same amount of vinegar. Put the cap on and shake the liquids up. Wait a couple of minutes and watch what happens. Now take the cap off and try to pour out some of the vinegar. When you tip the bottle up, only the oil pours out. Try putting your finger over the end of the bottle and turning it upside down – make sure you do this over a sink! What happens? The oil always floats on top of the vinegar so when the bottle is upside down, the vinegar drips out.

Cleaning water

EXPERIMENT

Filters are used to separate **mixtures** of solids and liquids. See how they work when you try to clean some dirty water. Mix some soil with water in a jug. Put a coffee filter paper, or some blotting paper, in a funnel (you can ask an adult to make a funnel by cutting off the top half of a plastic bottle and turning it upside down). Stand the funnel in a jar. Slowly pour the dirty water into the funnel. Wait for the water to drain away before adding a little more. The water that comes out in the jar should be much cleaner, because the filter catches the dirt and separates it from the water.

Ink test

EXPERIMENT

The ink in a felt-tip pen might look like it is just one colour, but it is actually made up of lots of different colours. See how many colours there are in each of your felt-tip pens by separating them out. Take a sheet of blotting paper and draw four different coloured spots or shapes about 3cm from the bottom of the paper. Dip the bottom of the.paper in a shallow dish of water, making sure the spots of ink remain above the water. The blotting paper soaks up the water so that it runs into the ink and separates the different colours. Which felt-tip pen was made up of the most colours?

Heat

We need heat to keep our bodies and houses warm and to cook our food. Like light and sound, heat is a type of **energy**. Heat flows from a hotter place to a colder place in different ways. The heat that you feel on your skin from the Sun travels by **radiation**. A teaspoon in a cup of tea gets hot by **conduction**. The hot air rising from the cup carries heat by **convection**.

straw
modelling clay
cork
bottle
water

Make a thermometer

EXPERIMENT

Fill a small bottle with cold water coloured with ink. Ask an adult to make a hole in a cork and push a straw through it. Use some modelling clay to seal any gaps around the straw. Fit the cork in the bottle and press down to force a little water up the straw. Put the thermometer in a warm place. Heat will flow to the water by **conduction**. As the water gets warmer, it **expands** (takes up more space) and rises up the straw. Now put the thermometer in a cool place. What happens to the water in the straw?

Coiled Snakes

Radiators heat a room mainly by **convection**. Like water, air **expands** when it is heated and takes up more space. The warm air rises above the cooler air (which takes up less space) and warms the room. Why not see for yourself?

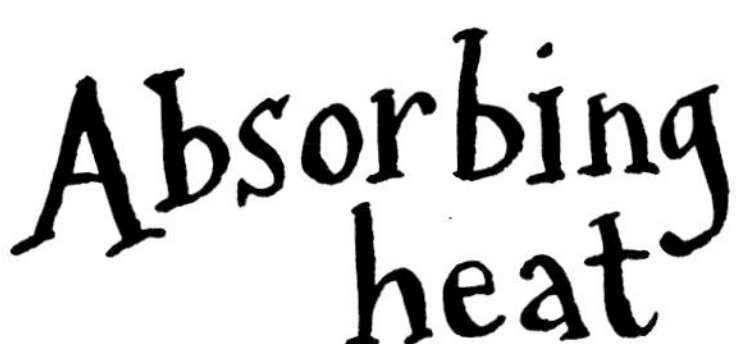

Absorbing heat

EXPERIMENT

Try wearing a black T-shirt and shorts for one hour on a hot day, then switch to a white T-shirt and shorts. In which T-shirt do you feel cooler? Darker colours are better at absorbing heat **radiation** from the Sun than light colours. If two objects made of the same substance but of different colours are placed near a heat source, the darker one feels hotter.

Which material is best?

EXPERIMENT

Insulation stops the **conduction** of heat and keeps things hot or cold. We use it in our homes to save energy and in winter we wear thick clothes that insulate us. Find out which materials make good insulators. Fill four jars with warm water to within 2cm of the top and put on the lids. Now wrap each one in a different material, such as leaves, nylon fabric, cotton wool, and feathers. Check the temperature of the water in the jars every five minutes. Which jar of water cools quickest? Which insulation material is best?

A vacuum flask keeps drinks hot or cold. The flask has two bottles, one inside the other, with very little air in between. This means heat cannot move by conduction or convection. The shiny walls on the inside of the flask cut down on heat radiation.

On a sunny day, birds and gliders can be seen soaring in the sky on rising convection currents of warm air, called thermals.

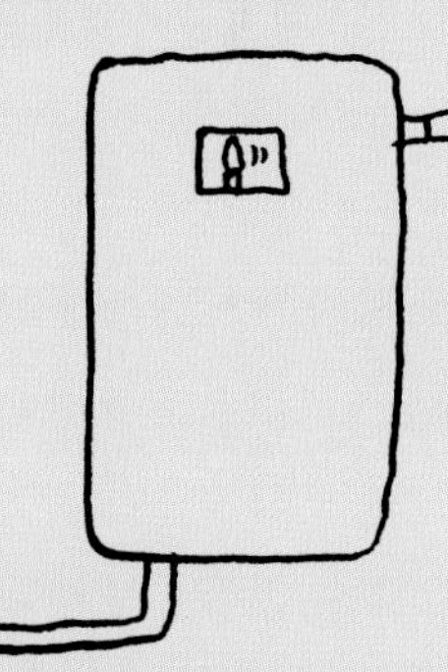

In a boiler, cold water moves along pipes which are heated by flames from oil or gas. The hot pipes pass their heat on to the water by conduction.

SCIENCE TIPS

There is a good reason why a space is left at the top of a drinks bottle. When the bottle gets warm, the drink inside it expands and takes up more room.

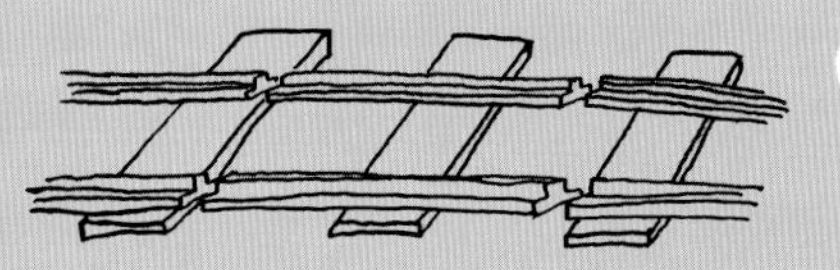

When railway tracks are built, small gaps are left to allow for expansion on hot days. Without the gaps, the railway tracks would buckle and break.

Expansion and contraction

Many materials get bigger (**expand**) when they are heated and get smaller (**contract**) when they are cooled. Liquids usually expand more than solids do when they are heated and gases expand most of all. When air gets hot, it rises above cooler air.

Make a hot-air balloon

flaps

EXPERIMENT

You will need: • a large piece of tissue paper • scissors • a felt-tip pen • glue • a hairdryer

1. Using the pen, carefully copy the shape shown on to the tissue paper and cut it out.

2. Fold the paper along the lines to make a rectangular bottomless box. Spread glue along the flaps and stick them down.

3. Ask an adult to hold your balloon while you blow hot air from a hairdryer inside it. Once it is full of hot air, the balloon will slowly rise to the ceiling. This is because as the heated air **expands**, it gets lighter too, so that the hot-air balloon rises.

1.

2.

3.

Problem unscrewing a lid?

EXPERIMENT

Next time you find it difficult to open a jar or a bottle, ask an adult to place it under a stream of hot water. Ask them to dry the lid and give it back to you. They'll be amazed when you manage to open it! The heat makes the metal lid **expand** so that it squeezes less against the bottle – and you can open it easily!

Pop!Pop!Pop!

When tiny kernels of corn are heated, a chemical reaction takes place and they expand with a sharp pop – making popcorn!

SCIENCE TIPS

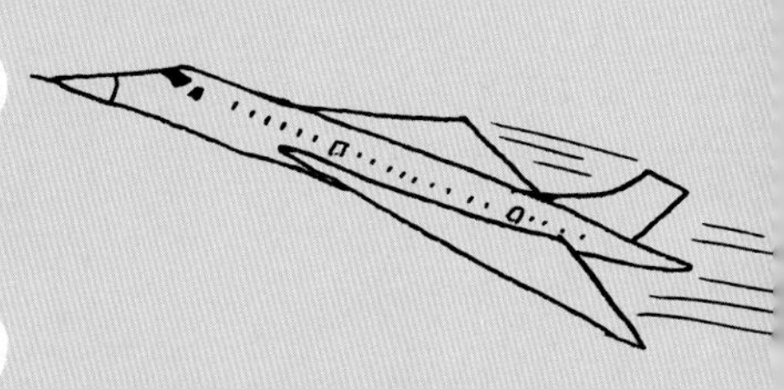

During flight, Concorde is half a metre longer than when it is on the ground! It flies so fast that the outside of the aircraft heats up to about 1,000°C, making the metal body expand.

A spin dryer has a thermostat to keep temperatures constant. The metal in the thermostat expands and bends when it gets too hot. When it bends it switches off the heater so that the dryer goes back to the right temperature.

SCIENCE TIPS

The surface of the water in ponds and lakes is strong enough to support the weight of small animals and bugs such as the pond skater.

Air and water

Air is all around us and water is one of the most common substances on Earth – it covers nearly three-quarters of the planet's surface. Water exists naturally as a gas (water **vapour** in the atmosphere), a liquid (in oceans, lakes and rivers) and as a solid (ice and snow).

Make a water wheel

EXPERIMENT

You will need: • a straw about 7cm long • a round plastic lid • scissors • a knitting needle

1. Take the plastic lid and ask an adult to make a hole in the middle big enough for the straw. Using the scissors, carefully cut slits round the edge of the lid. Bend the flaps of plastic to the left slightly, so that they overlap.

2. Put the straw through the hole and then put the knitting needle through the straw.

3. Hold the water wheel under a running cold tap. What happens?

1.

2.

3.

On the surface

EXPERIMENT

When you look at water dripping from a tap, it looks as though the water drops have a skin. This is caused by a force called surface tension, which pulls the surface of the water tight. Find out how strong this 'skin' is by filling a small bowl with water and gently placing a paper-clip on the surface so that it floats. Now add a few drops of washing-up liquid to the water. What happens to the paper-clip? The washing-up liquid weakens the surface tension.

SCIENCE TIPS

When you take the plug out of the bath, the water is sucked down the plughole. If you put your hand over the plughole, you can feel the water being tugged. As it is sucked down, it spins into a whirlpool.

Lawn sprinklers are used to water gardens and parks when there has been only a little rain. They spray water out in high-speed jets. The faster the flow of water from the tap, the faster the sprinkler turns.

What's the time?

EXPERIMENT

Carefully make a small hole in the bottom of a paper cup with a drawing pin. Attach the cup to the top of a ruler with some tape. Tape another cup, without a hole, underneath the first cup. Stand the ruler upright (you may need to secure it with some modelling clay). Cover the hole in the top cup with your finger while you fill it with water. Take your finger away and watch the water drip into the bottom cup. Time the drips with your watch and mark the level of the water in the lower pot every minute and every five minutes. Every time you fill the top cup, you will be able to tell the time by how much water has dripped into the bottom cup.

SCIENCE TIPS

Many traditional windmills have been replaced by wind turbines. Their blades are specially designed to catch as much wind as possible. Wind farms are built where it is windy all year round.

Different birds have different kinds of wings. Hawks and swallows have swept back, narrow wings for speed, while vultures have large, broad wings for gliding.

Using air

Without air we could not live. We cannot see it or smell it, but we can feel it when it moves. Just like water, air has many uses. Sailing boats have large sails which catch the wind to push them through the water. Windmills harness the power of the wind to grind wheat into flour or make electricity.

Build a windmill

EXPERIMENT

You will need: • a square piece of wrapping paper • a drawing pin • a straw • a small bead • glue • scissors • a pencil • modelling clay

1. Fold the paper in half and in half again to find the middle. Draw four wavy lines and five holes, as shown, on the side of the paper without the pattern. Ask an adult to cut out the holes and then cut along each of the lines.

2. Fold in each of the four corners so that the holes meet in the middle of the square.

3. Glue the corners over the centre hole. When the glue has dried, carefully push the drawing pin through the middle of the windmill. Thread a bead onto the pointed end of the pin and then push it into the straw. covering the end with modelling clay. You may need an adult to help you.

4. Hold your windmill in the wind.

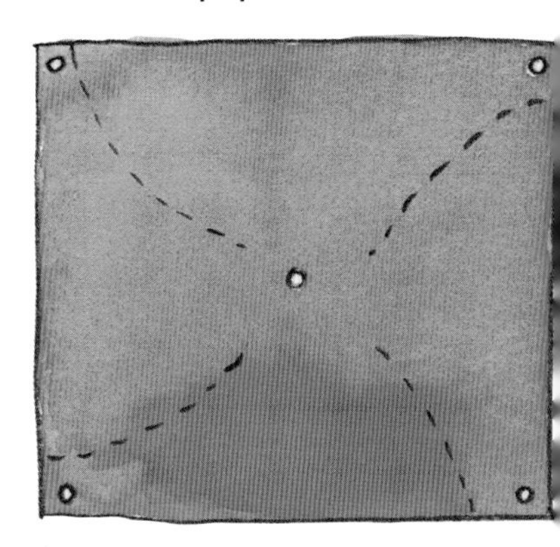

1.

4.

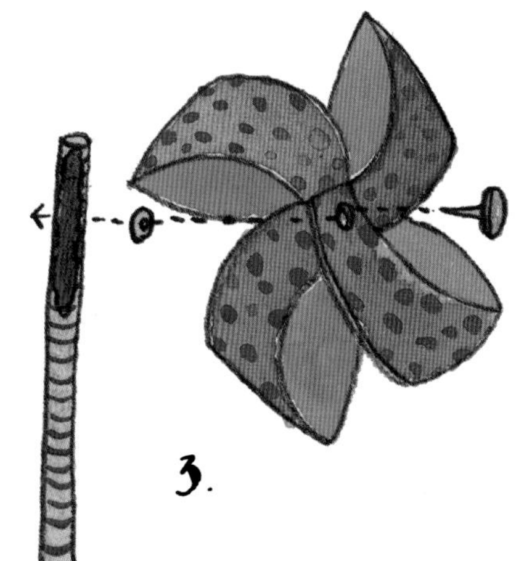

3.

2.

Paper planes

When air rushes over the wings of a plane, it lifts the plane and keeps it in the air. Make your own plane and see how far it will fly. Try making the wings a different shape. How far will it fly now?

1. Fold the paper in half and make a sharp crease.

2. Open the paper out and fold in the two top corners.

3. Fold the sides into the middle again to get a pointed shape.

4. Lift the paper up and fold the two top edges down.

5. Open out the wings and tape them together.

6. Give the plane a little push and watch it fly!

Catching the wind

matchbox

modelling clay

toothpick

sail

Take a small rectangular piece of paper and colour it in the pattern you would like for your sail. Make a small hole in the top and bottom of the sail with a toothpick. Put the toothpick through the holes to give the sail a billowy shape and fix it to the inside of an empty matchbox using some modelling clay. Float your boat on some water and blow on the sail to make it move. What happens if you make a larger sail for your boat?

SCIENCE TIPS

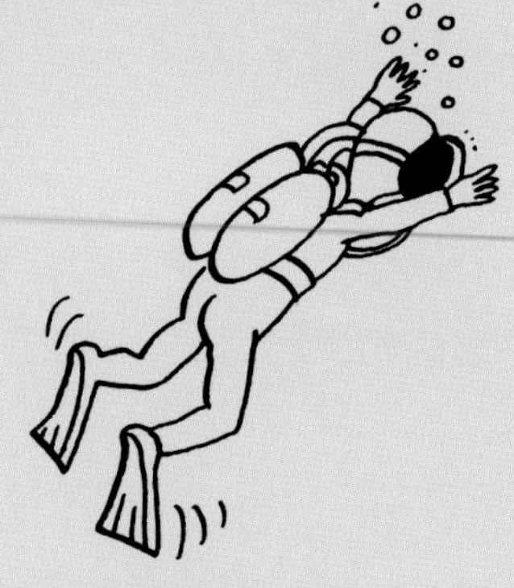

Divers use cylinders of squashed air to help them breathe underwater. More air can be stored in the cylinders if it is squashed in, so that the divers can stay underwater for longer.

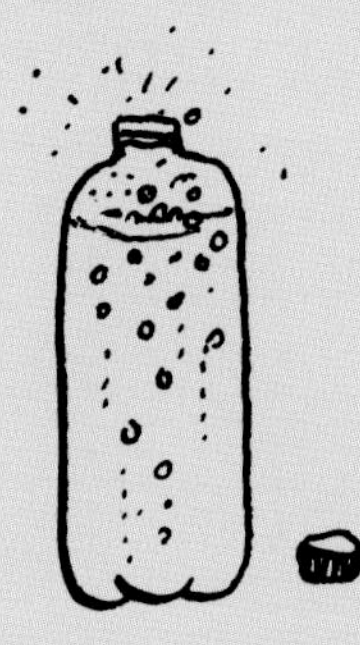

The 'fizz' in fizzy drinks comes from a gas called carbon dioxide which is forced into the liquid under pressure. When the bottle is shaken, the pressure increases and when the cap is taken off the squashed gas flies out.

Under pressure

Although you cannot feel it, the air around you is constantly pressing down on your body. You can feel air **pressure** by pressing against an inflated balloon. The air that has been squeezed into the balloon pushes back at you. When air is squashed into tyres, it can push back enough to support the weight of cars and even lorries.

That sinking feeling

plastic bottle

bowl

pen top

modelling clay

EXPERIMENT

Find out how a submarine dives to the bottom of the ocean using a pen top, modelling clay and a bottle of water.

1. Wrap some modelling clay round the bottom of the pen top to weigh it down, making sure you don't block the hole. If there is a hole in the top of the lid, seal it with the clay.

2. Put the pen top in a bowl of water and add or remove clay until it floats just below the surface.

3. Fill the bottle with water and drop the pen top inside. Screw the cap on and squeeze the sides of the bottle. What happens? By squeezing the bottle you push water inside the pen top and squash the air bubble trapped inside it. The pen top weighs more than before (because it has more water in it) and sinks. When you stop squeezing the bottle the bubble in the pen top grows again, making it lighter so that it rises. A submarine dives and surfaces in the same way.

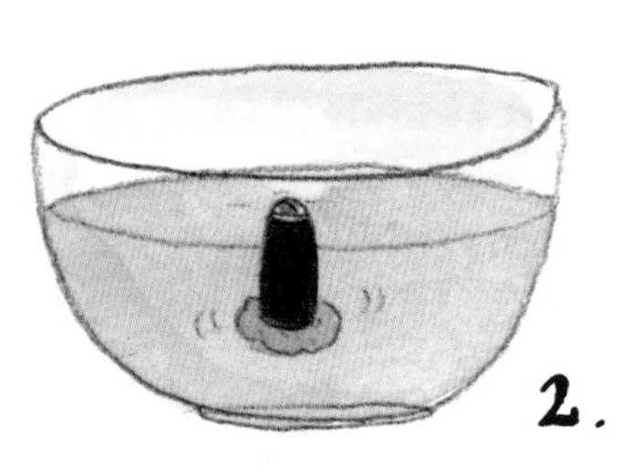

Make a Submarine

EXPERIMENT

This experiment will also show you how a submarine works. Why not try it in the bath? Fill a plastic drinks bottle with water and watch it sink to the bottom of the bath. Now put one end of a bendy straw in the bottle and blow air into it. The bottle becomes lighter because it has less water in it and more air. A real submarine has special tanks which it fills with seawater to make it dive. To rise, squashed air carried inside the submarine is pumped into the tanks, and the seawater is pumped out.

A kind of magic

EXPERIMENT

Fill a glass right to the top with water and slide a postcard over the top. Hold your hand on the card and quickly turn the glass upside down – make sure you do this over a sink! Slowly take your hand away. What happens? The postcard does not fall off because the **pressure** of the air pushing up on the postcard is more than the pressure of the water pressing down on the postcard.

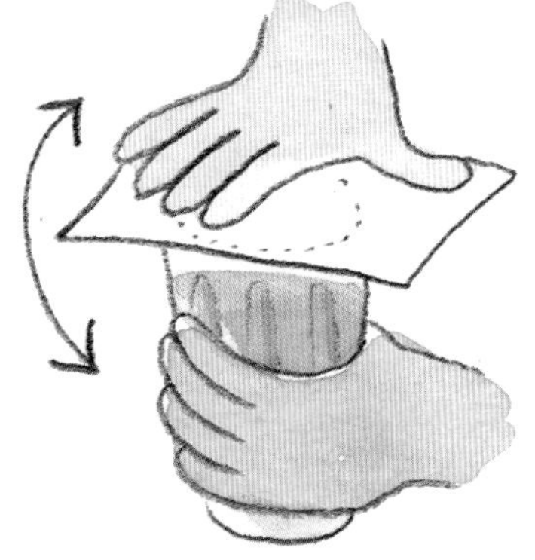

Forces

A **force** is a push or a pull – whenever you push a car or pull a thread you are applying a force. Forces exist in pairs. When one force pushes on something, another force pushes back. When you stand on the floor your weight pushes down but the floor pushes back at the same time – otherwise you would fall through the floor!

SCIENCE TIPS

Every time you push on a light switch or open a can of drink, you are using force.

You can start a swing by pushing off the ground with your feet. The force sets the swing in motion and once you have started the swing, it is easy to keep it going.

An object cannot begin to move on its own – it needs a force to make it move. A force is applied when you kick a football or hit a tennis ball with a racket.

EXPERIMENT

Lean against a wall with your arms stretched out and your feet slightly apart. Ask a friend to lean on you by putting their hands on your shoulders. Ask more of your friends to join in. You should be able to support all your friends! As you push on the wall, the wall pushes back with equal **force**, keeping you upright and as your friend pushes on your shoulders, you push back, and so on.

Make a paddle boat

You will need: • an empty matchbox
• 2 toothpicks • an elastic band • card • scissors

1. Carefully push each of the toothpicks into the sides of the matchbox so that they are firmly in place and pointing down slightly.

2. Slide an elastic band over the toothpicks.

3. Cut out a piece of card about the size of the end of the matchbox. Slide it into the elastic band and wind it up. Holding the card in position, put the model in a large bowl of water or in the bath.

4. Let go of the piece of card. As the paddle spins it pushes the water backwards. Because forces exist in pairs, the water pushes back on the paddle, making the boat move forward.

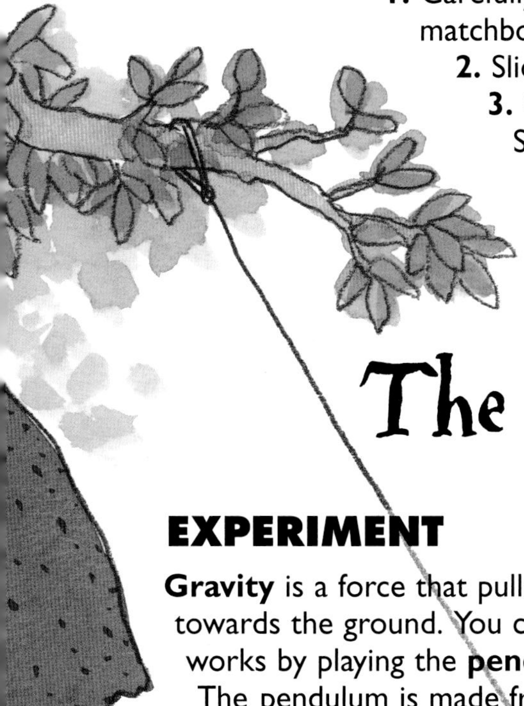

The pendulum game

EXPERIMENT

Gravity is a force that pulls objects towards the ground. You can see how it works by playing the **pendulum** game. The pendulum is made from a piece of string and a tennis ball. First make some skittles by putting pencils in cotton reels. Set up the skittles on a tray underneath a tree. Now attach a long piece of string to a tennis ball using strong sticky tape. Ask an adult to tie the other end of the string to a low branch. Make sure the ball just hits the tops of the pencils when it is swung. Each time you let go of the pendulum, the force of gravity pulls the ball towards the ground so that it swings. The aim of the game is to swing the pendulum so that it pushes over the skittles.

SCIENCE TIPS

Scissors act as a pair of levers to help us cut paper. Spanners help us to turn nuts, and pliers help us to pull nails out of a wall.

A wheelbarrow is a type of lever that helps us to carry heavy things more easily.

A crane uses a complex pulley system to move heavy objects around a building site.

Simple machines

We use simple machines to make difficult jobs easier. **Levers**, **pulleys** and **gears** are all types of simple machine. A seesaw is a lever – it helps lift us into the air. A pulley makes it much easier to pull a bucket up from the bottom of a well and a bicycle has gears to make it easier for us to pedal up and down hills.

Counting coins

EXPERIMENT

Levers make it easier to lift things. Make a seesaw by putting a pencil under the middle of a 30cm ruler. Place 10 coins halfway between one end of the ruler and the pencil. Now start putting more coins on the end of the ruler furthest from the pile. How many coins do you have to put on the ruler to lift the pile of 10 coins?
It takes fewer coins because the ruler acts as a lever, magnifying the weight of the coins further away from the pencil. What happens if you put the smaller pile of coins nearer the pencil? Will they still lift the pile of 10 coins? Move the piles of coins around to find out how the lever works best.

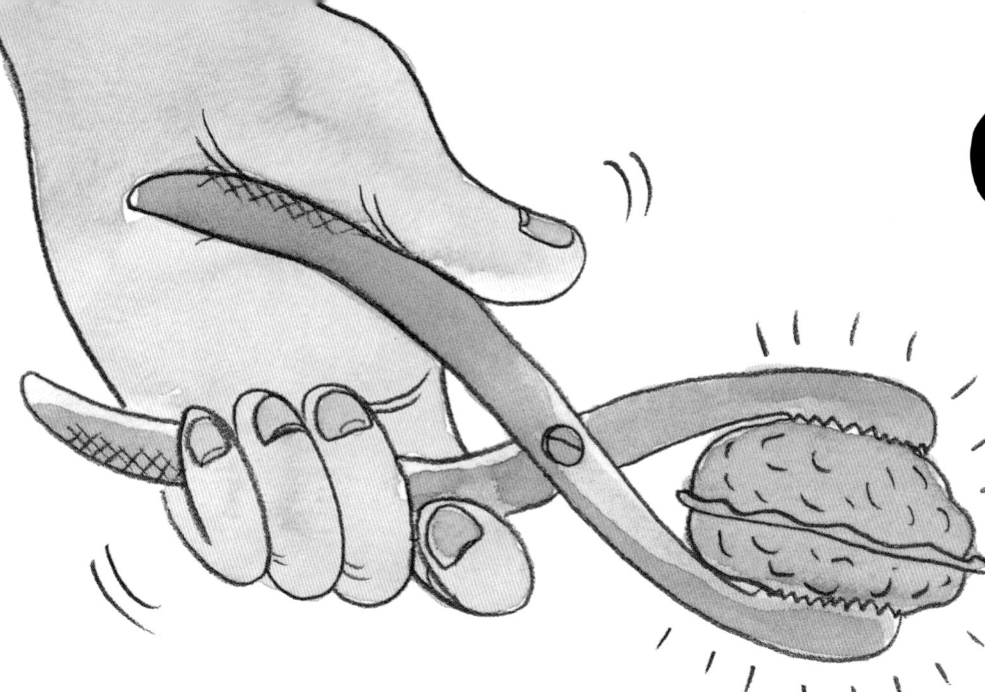

Cracking nuts!

Take a walnut and put it in your hand. Make your hand into a fist and use all your strength to try to crack the shell. What happens? Now try to crack the walnut with a nutcracker. It is much easier because the nutcracker acts as a **lever** and magnifies the push you apply at the handles.

Working out gears

EXPERIMENT

Gears are wheels with teeth that slot together. A bicycle has a set of gears to make pedalling easier. Rest a bicycle upside down and mark a spot on the back tyre with some coloured tape. Turn the pedals very slowly with your hands. As soon as the pedals have gone round once, use the brakes to stop the wheel. How far has the wheel turned? Change gear and repeat the experiment. How far has the wheel turned this time?

Friction

Friction is a force that tries to slow things down or stop things moving. It is produced when two objects touch each other. When you rub your hands together, the friction makes them feel warmer. Friction can be good – without it, our feet would always be sliding around. But there are other times when friction is a bad thing, for instance when you cycle into the wind.

A slippery job

Wash your hands in soapy water without rinsing or drying them. Now try to unscrew the lid of a jar. Can you get the lid off? Rinse your hands and dry them thoroughly. What happens when you try to take the lid off this time?

The great friction challenge

EXPERIMENT

Put a piece of wood, a stone, an ice cube, a pencil eraser and a key on one end of a wooden chopping board. Slowly lift up this end of the board to make a ramp. How high do you have to lift it before the objects start to move? Which object moves first? The greater the **friction** each object has to overcome, the more the board has to be tilted. What happens if you put the same objects on a metal tray? Write the results of the experiment in your notebook.

SCIENCE TIPS

The bottom of a bath can be very slippery but a rubber bath mat creates more friction, to give a better grip.

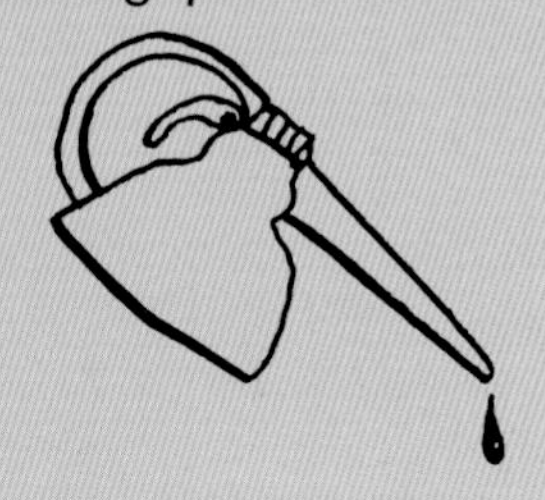

People oil machines to reduce friction between the metal parts, making them work better.

Some shoes have soles with tread to give a good grip on the ground so that we don't slip over.

Make a balloon hovercraft

A hovercraft can move much faster than a ship at sea because it rides on a cushion of air. The **friction** between the hovercraft and the air is less than the friction between the ship and the water. Make your own hovercraft and see how easily it moves on a cushion of air.

EXPERIMENT

1. Ask an adult to cut the top off a plastic drinks bottle, as shown. Get them to make a small hole in the bottle cap.

2. Take the balloon and blow it up. When you have finished, pinch the neck of the balloon with your finger and thumb so that no air escapes.

3. While pinching the neck of the balloon, put it over the bottle cap. You may need somebody to help you with this.

4. Stand your model on a smooth surface, such as a table, and let go of the balloon. Give the bottle a little push. The air rushes out, pushing the bottle neck off the ground so that it glides like a hovercraft.

SCIENCE TIPS

You can see sunlight on a sunny day when it pours through a window. Dust scatters some of the light towards our eyes, making the sunbeams visible – in the same way that a car's headlight beams show up in fog.

Light

The only way we can see an object is if rays of light hit it and reach our eyes. That's why we can't see things in the dark. The objects are still there but there is no light to reveal them. Light is a kind of **energy** that travels in waves. It can pass through **transparent** objects like glass but cannot pass through **opaque** objects like wood.

Ghostly shadows

Because light cannot travel through an **opaque** object, a shadow forms behind the object. That's why you have a shadow on a sunny day – because your body blocks the sunlight. Find out more with this spooky experiment! Copy this ghostly shape on to a piece of stiff white card. Cut out the shape and stick it to a ruler. Turn on a lamp and turn out the main lights so most of the room is dark. Hold up the mask between the lamp and the wall. You should see a ghostly shadow. What happens to the shadow if you move the mask nearer the lamp? What happens if you move it further away?

Bending light

EXPERIMENT

Light can produce some very strange effects – and play tricks on your eyes! Half fill a glass with water. Put a pencil or a straw in the water and look at it from the top, bottom and sides. What happens when you look at it through the sides of the glass? The reason the pencil looks bent is that light travels slower through water than air. As the light enters the glass of water, and as it leaves the glass, it changes speed and direction – making the pencil look bent!

Make a sundial

EXPERIMENT

During the day, the Sun changes its position in the sky and moves from east to west. This means the shadows it casts move too and by looking at them at different times of day, you can tell what time it is. On a sunny day, place a large piece of paper or stiff white card in a spot outside where there is no shade. Turn a small flower pot upside down in the centre of the card. Stand a pencil or stick in the hole in the base of the pot. Every hour, mark the position of the shadow the stick makes on the paper, using a marker pen. Write the time by the shadow lines. Now you will be able to tell the time on a sunny day by looking at where the shadow is on your sundial.

SCIENCE TIPS

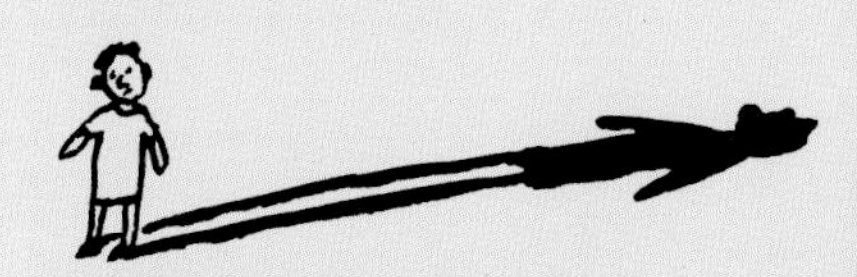

Shadows vary in length throughout the day. Early in the morning and late in the afternoon the Sun is low in the sky and the shadows are long.

At midday, the Sun is very high in the sky and shadows are shorter.

An eclipse takes place when the Sun is blocked by either the Moon or the Earth. When the Moon lies between the Earth and the Sun, it blocks the Sun and its shadow falls on the Earth, creating darkness during the day. This is called a solar eclipse. A lunar eclipse occurs when the Earth is between the Sun and the Moon and its shadow falls on the Moon.

SCIENCE TIPS

Dentists use concave mirrors to make teeth look bigger so they can examine them more easily. Concave mirrors are also used for shaving because they make the face look bigger.

All about reflections

When light hits a surface, some of it bounces off, or is **reflected**. Mirrors are very shiny surfaces designed to reflect nearly all the light that hits them. When you look in a flat mirror, you see a reflection of yourself which is the same size as you but back to front. When you look in a curved mirror your reflection is usually a different shape and size to you!

Shiny spoons

EXPERIMENT

The surface on the *inside* of a spoon is **concave** – it bends in, like a cave. The surface on the *outside* of a spoon is **convex** – it bends out. First look at your reflection on the inside of a shiny spoon and then look at it on the outside. What differences can you see? The crazy mirrors that you get at funfairs bend in all sorts of ways so that you look big, small, fat, thin and even wavy!

Back to front

EXPERIMENT

See if you can write your name back to front on a piece of paper so that it reads correctly when you hold it up to a mirror. Now place the mirror in front of you and try writing your name the right way around while looking at what you are doing in the mirror!

Seeing around corners

you need:

EXPERIMENT

A submarine has a periscope so its crew can see above water when the submarine is submerged.

1. Fold a square piece of card diagonally to make a right-angled triangle. Make sure it has two sides of equal length.

2. Use the triangle to draw two diagonal lines on the side of an old milk carton. Do the same on the opposite side of the carton. Ask an adult to cut along the lines to form slits. Take a mirror with its shiny side facing down and slide it into the top slit. Push it through so it appears on the other side. Push a second mirror through the bottom slit, making sure it is facing up.

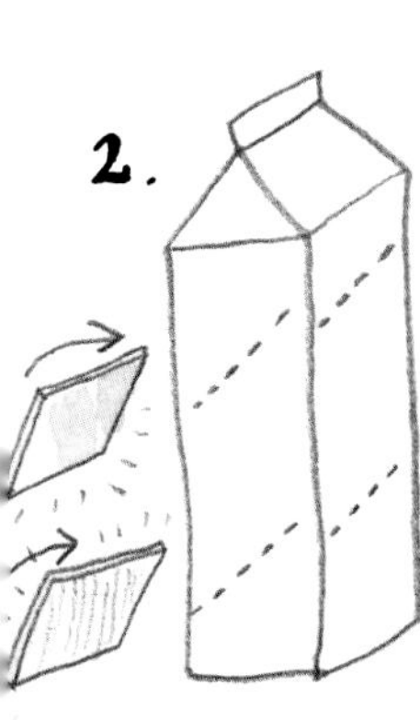

3. Make a small hole in the back of the carton and cut a large square from the front, as shown.

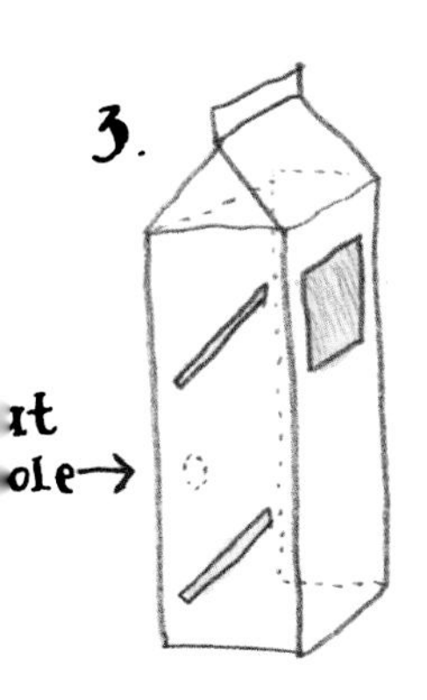

4. Hold the periscope in front of a wall or a hedge so it pokes over the top and look through the hole. What can you see?

4.

Colour

Sunlight does not look as if it has any colour, so it is called white light. But sunlight is actually made up of lots of different colours. You can see these colours when it is sunny and raining at the same time, as a rainbow forms. The raindrops split the light into seven different colours – red, orange, yellow, green, blue, indigo and violet.

Amazing glasses

You'll need:

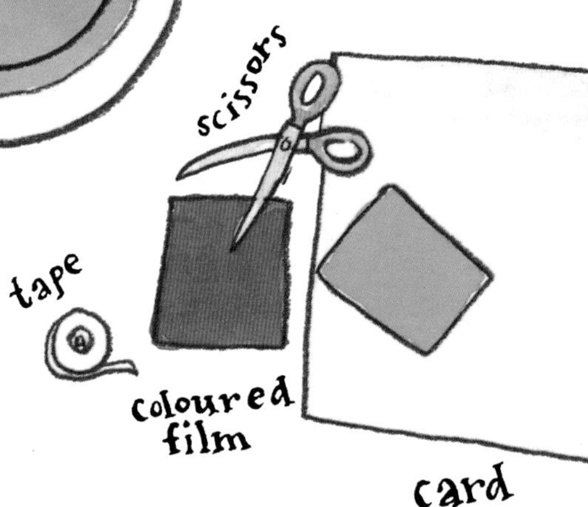

EXPERIMENT

Copy the glasses frame on to a piece of card and draw two holes to look through. Cut the shape out and ask an adult to help you cut out the eye-pieces. Tape some green coloured film over one eye-piece and some red film over the other. Bend the arms of the glasses back and put the glasses on. Have a look around. Does everything look different? This is because the red and green film stops certain colours of light getting through.

Wow!

Make a rainbow

EXPERIMENT

On a sunny day, you can create your own rainbow. Put some water in a shallow dish and prop up a small mirror in the water at an angle. Place the dish near a window and position the mirror so that sunlight hits it. The light passes through the water and bounces off the mirror, making a faint rainbow appear on the wall. If you do not have white walls, take a large piece of white card and hold it in front of the wall. Adjust it until you can see the rainbow.

In a spin

EXPERIMENT

Stand a glass on a piece of white card and draw round the base. Carefully cut the circle out. Divide the circle into three, equal-sized sections and colour one in blue, one in red and one in green. Ask an adult to make a hole in the centre of the card (using a hole punch), so that you can fit a pencil through it. When the card is spun on the pencil tip, what colour do you see? Make lots of spinning discs using different combinations of colours. What colours do you see when you spin them?

SCIENCE TIPS

Special pictures called holograms are often printed on credit cards. The colours change as you look at them from different angles as the angle of the light that strikes the hologram and enters your eye changes.

Glass is not completely transparent. When light hits the glass at certain angles it is reflected, which is why you can see your face in a train window.

Sound

When something **vibrates**, or shakes very, very quickly, it makes the air around it vibrate too. When these vibrations travel to our ears, we hear a sound. Like light, vibrations can be **reflected** (bounced back). In a mountain valley you will often hear an echo, which is caused by a sound being reflected off a faraway surface.

This is fun!

Bouncing sounds

EXPERIMENT

Open two umbrellas and put them on the ground facing each other so their handles meet. Prop the umbrella handles up on some books. Fix a ticking watch to the shaft of one of the umbrellas. Put your head inside the other umbrella and rest your ear on the umbrella shaft. The ticking should sound as if the watch is right next to your ear! This is because the sound of the ticking bounces off the inside of one umbrella towards the other.

See the vibrations

EXPERIMENT

Sound is invisible, but you can see and feel the way it makes the air **vibrate**. Stretch some clingfilm over a bowl until it is tight and sprinkle some sugar on to it. Hold a baking tray over the bowl and bang it hard with a wooden spoon. The vibrations made by banging the tray make the air vibrate and the sound is carried to the bowl, making the clingfilm vibrate and the rice jump.

SCIENCE TIPS

The sound made by musicians playing in a concert hall is reflected by specially designed surfaces that control the sound to make it as good as possible.

Bats use echoes to 'see' at night. When they send out a sound, it bounces off an object such as a moth and comes back to them so that they can figure out where the moth is.

Making a long-distance call

EXPERIMENT

Ask an adult to make a hole in the bottom of two plastic cups or yoghurt pots. Thread one end of some string (up to about 15 metres long) through one cup and tie a knot inside the cup. Thread the other end of the string through the second cup and tie another knot. Hold the cups so that the string is tight and get a friend to whisper into one cup while you listen with the other. The sound of their voice makes the string **vibrate**, carrying the sound to your ear.

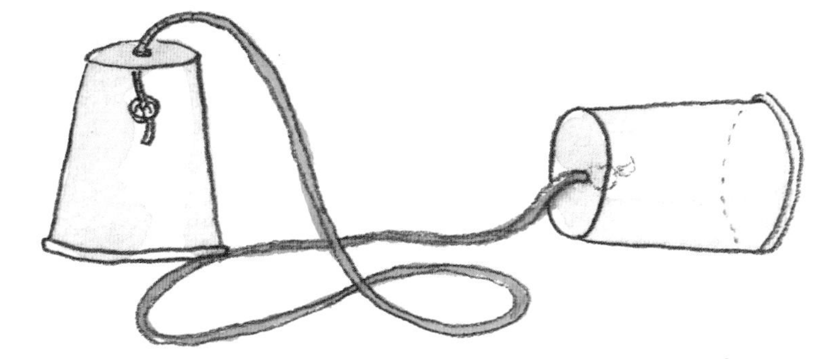

SCIENCE TIPS

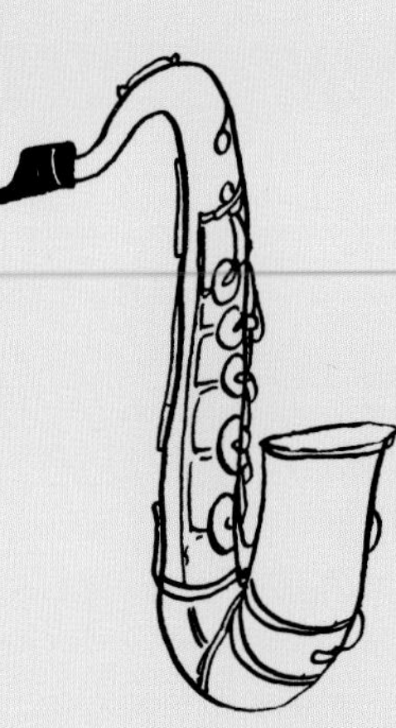

The note produced by a brass instrument depends on the length of the pipe. The longer the pipe, the lower the note.

In a stringed instrument, the note depends on the size, length and tightness of the string.

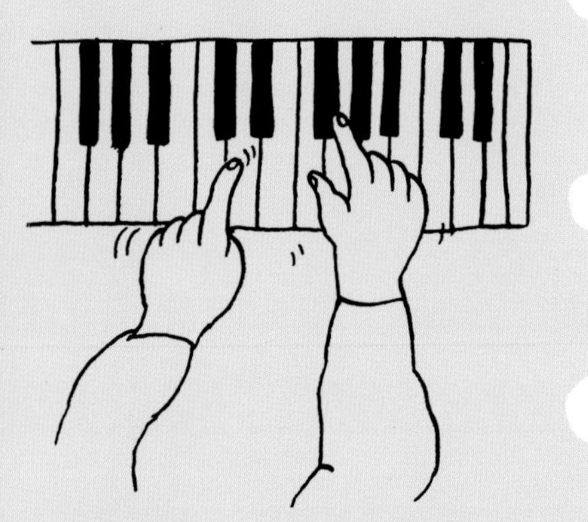

When a piano key is pressed, a hammer hits a wire inside the piano. The wire vibrates and produces a sound.

Musical notes!

Musical instruments can produce lots of different sounds. Each type of instrument produces sound from **vibrations**. The sound can come from a plucked string (in stringed instruments), from blowing into a pipe (in brass and wind instruments), or from hitting hard surfaces (in percussion instruments).

Play the guitar

EXPERIMENT

The guitar, like the violin and the double bass, is a stringed instrument. Guitarists produce sounds by plucking the strings. Why not see for yourself? Stretch some elastic bands of different thickness over a rectangular plastic box (such as an empty margarine tub). Listen to the sound that is produced when you pluck each of the bands in turn. When the elastic bands **vibrate**, they make the air in the box vibrate. This makes the sounds louder. Thinner bands produce a different sound from the thicker ones. Which make the highest sounds?

A bottle orchestra

EXPERIMENT

Put some plastic bottles in a row. Add a little water to the first bottle and then increase the amount of water in the bottles until the last one is nearly full. Blow across the top of each bottle to make a sound. Compare the sounds made by the different bottles. The more water in the bottle, the higher the note.

Make a whistle

1.

2.

3.

4.

1. Take a paper straw and squeeze one end flat.
2. Carefully cut the end into a triangular shape.
3. Cut the sharp end of the triangle off.
4. Ask an adult to help you cut three or four small holes in the straw, like a whistle.
5. Put the pointed end of the straw into your mouth and blow. You should hear a sound. Take it in turns to cover each of the holes with your fingers. How many different sounds can you make?

5.

The Doppler effect

EXPERIMENT

Ask a friend to cycle past you while sounding their bicycle horn or singing a steady note. Listen carefully to the sound and how it changes. As your friend moves towards you, the note gets a little higher. As they cycle away from you, the note gets lower. The same thing happens when you hear police sirens. This is called the Doppler effect.

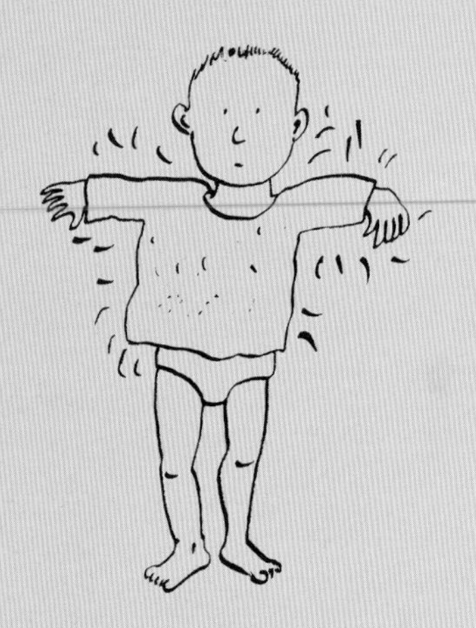

Clothes made from artificial fibres like nylon often build up static during the day. When you take them off at night, you can feel and hear the crackle of the static. If you turn off the light and look in a mirror, you may even see sparks.

Electricity

Many objects that we use every day are powered by electricity – from computers and hairdryers to lamps and washing machines. There are two types of electricity. **Static electricity** stays in one place and **current electricity** moves through things. Static electricity is produced when some materials are rubbed together. It causes crackles when you comb your hair and makes dust stick to television screens.

Jumping frogs

EXPERIMENT

Take a sheet of green paper and tear it up into small pieces. Use a pen to add two black dots to each piece of paper for your frogs' eyes. Put the frogs in a pile on a table. Now comb your hair with a plastic comb several times. Hold the comb over the frogs and they should come alive! The frogs jump up because the comb has **static electricity**.

Sticky balloons

Take a balloon that has been blown up and rub it against a nylon jumper. Now try sticking the balloon to the wall. It doesn't fall off because it is charged with **static electricity** and is attracted to the wall.

SCIENCE TIPS

Lightning is caused by a natural build-up of static electricity in clouds. The lightning strike is just a giant spark of electricity.

Some cars have a thin rubber strip at the back which hangs down to the road surface. This stops static building up as the car is driven along the road.

Attraction and repulsion

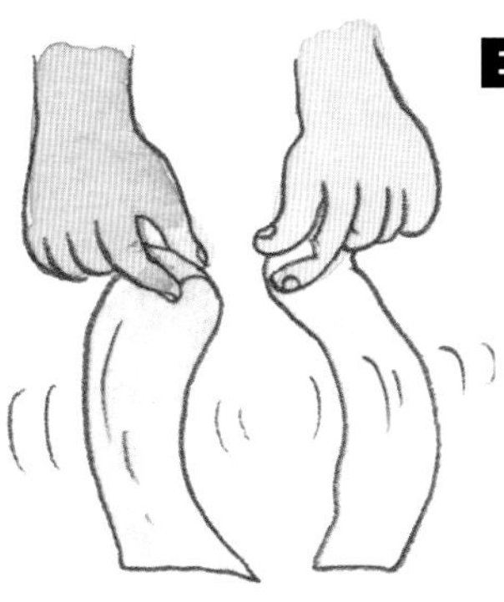

EXPERIMENT

Take two strips of clingfilm and run them through your fingers. The static on the clingfilm will make the strips cling to your fingers. Try holding the strips out and bringing them close together. What happens? Objects that have **static electricity** aren't always attracted to things – sometimes they are repulsed or repelled (pushed away). The clingfilm pieces push each other apart.

Bending water

EXPERIMENT

Take a plastic comb and rub it against a woollen jumper or scarf. Now turn on a tap so that a trickle of cold water comes out. Place the comb near the water. What happens? Is the water attracted to the comb or is it repelled?

Moving electricity

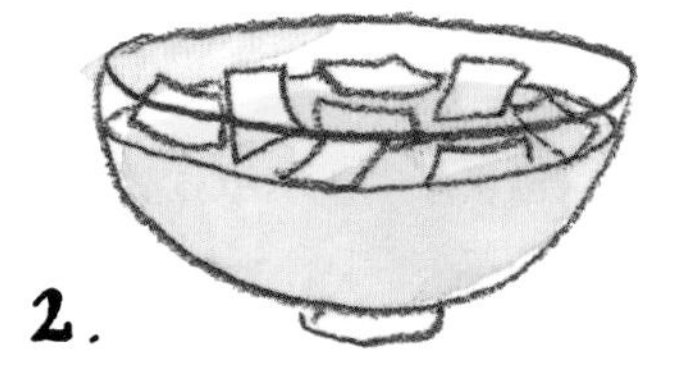

Objects that need **current electricity** (moving electricity) are powered by **batteries** or by electricity which travels along wires from a power station. The electricity will only flow if a complete **circuit**, or circular path, can be made from the wires. The circuit is completed by a switch, which turns the appliance on. When the switch is turned off, the circuit is broken and the appliance is turned off.

3.

What a shock!

You will need: • a paper towel • a lemon • 5 copper coins • 5 silver coins • a small bowl • scissors

1. Cut the paper towel into nine square strips, each measuring about 2cm.
2. Squeeze some lemon juice into the bowl and soak the pieces of paper in the juice.
3. Build a tower of money, alternating between the copper and the silver coins. Put a piece of paper between each coin, finishing with a coin on each end of the pile. Carefully turn the pile on its side.
4. Wet one finger of each hand and carefully lift up the pile. What happens? The metal in the coins and the acid of the lemon act like a **battery**. When you touch each end of the pile, you are completing a **circuit** and should get a tiny electric shock!

4.

Make a lighthouse

A **battery** has chemicals inside it. These produce electricity that flows when the two ends of the battery are connected with wire. Lemon juice is an acid that can be used to make a battery. Before trying this experiment, squeeze the lemon to free up the juices. If the experiment does not work, try using a 4.5V battery instead of the lemon.

EXPERIMENT

You will need: • an empty milk carton • several sheets of paper • paints • 2 pieces of copper wire (from a hardware store) • 1.5V torch bulb • a bulb holder • a lemon • a paper-clip • a brass drawing pin • modelling clay • scissors

1. Take the milk carton and ask an adult to cut off the top.

2. To make the lighthouse building, turn the milk carton upside-down and paint it with red and white stripes. Use the modelling clay to make rocks to go around the base. Paint the sheets of paper blue and fold them like a fan for the waves.

3. Connect the ends of the two pieces of copper wire to the bulb and stand it in a bulb holder on top of the lighthouse. Cut the lemon in half and make two slits in the lemon peel. Push the paper-clip into one and the drawing pin into the other. They should be close together but not touching. Tie the end of one of the copper wires around the paper-clip and the other around the drawing pin.

4. The lemon acts as a battery and produces electricity, turning the light on so that ships at sea don't hit the rocks!

SCIENCE TIPS

Hydroelectric dams turn the energy of falling water into electricity. The electricity is then sent along wires to people's homes and businesses.

Some magnets are shaped like horseshoes and some are like bars, while fridge magnets come in all shapes. Use a magnet to find magnetic things in your home.

Many everyday electrical appliances use magnetism – including telephones, televisions and computers. Never place a magnet near these machines or they may stop working.

Magnets

Magnets have a special power which enables them to attract things made from iron or steel. One end of a magnet is called the north pole and the other end is called the south pole. If you bring two north poles together they repel each other, or push each other away. But if you put a south pole next to a north pole, they jump together because opposite poles attract.

Make a compass

Believe it or not, the Earth is a giant magnet with a north and south pole. That is why a compass needle always points north. Make a temporary magnet by stroking one pole of a magnet along a sewing needle from top to bottom, at least 20 times. Lie the needle on a slice of cork and float the cork on some water in a small bowl or saucer. The needle swings round so that one end points north and the other end points south.

The magnet game

Draw a maze on the top side of a thin paper plate. Put a paper-clip on the plate at the start of the maze. Hold the plate with one hand and with the other, hold a magnet under the paper-clip. The aim of the game is to lead the paper-clip through the maze with the magnet. Get a friend to time you and then see how long it takes them.

Fishing for treasure

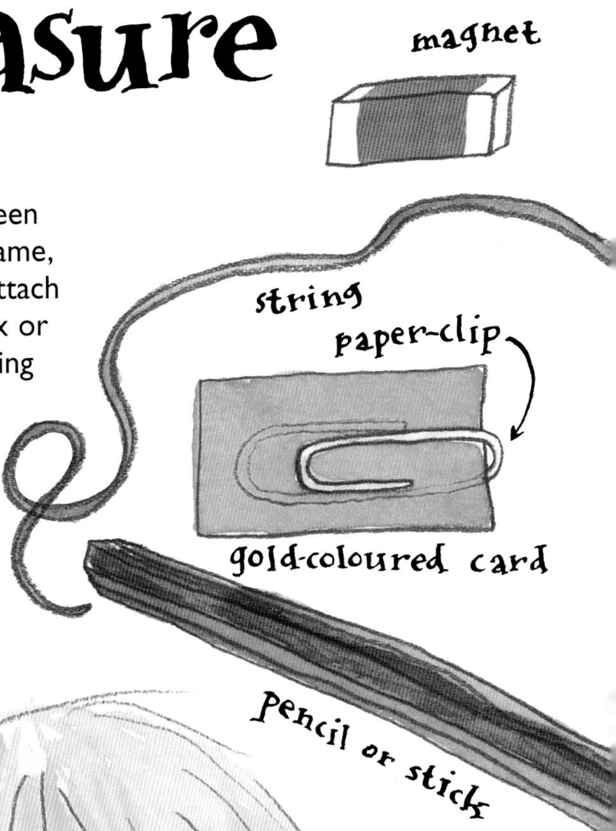

EXPERIMENT

At the bottom of a canyon, there is a pile of gold bars that has been abandoned by robbers. How many can you rescue? To play this game, you need to cut out 12 small rectangles of gold-coloured card. Attach a paper-clip to each gold bar and put them in an empty tissue box or shoe box. Each player needs a pencil or a stick. Tie a piece of string around the pencil and stick it down with some sticky tape. Tie the other piece of string around a magnet and again secure it with sticky tape. Now each player is ready to start fishing for gold. The one who picks up the most gold bars with their fishing magnet wins!

magnet

string

paper-clip

gold-coloured card

pencil or stick

Science Party

Now that you've mastered the Science School experiments, why not throw a party for your friends? Visit the library or look on the Internet to find out about the world's great scientists and get everyone to dress up. You could be Louis Pasteur, Archimedes, Marie Curie or even Albert Einstein!

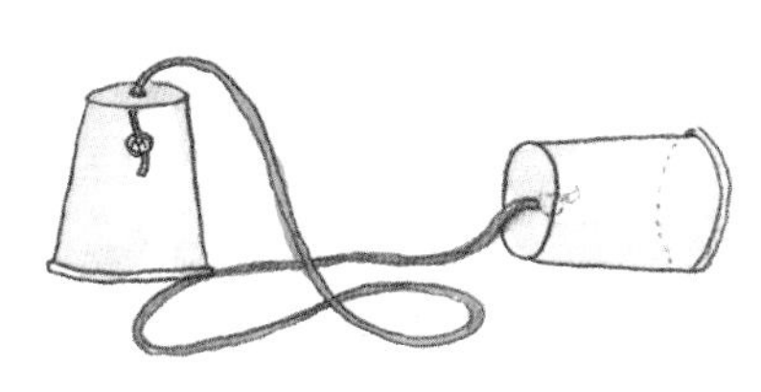

Glossary

Acid
A type of substance that has a sour taste, for example vinegar. Strong acids can be dangerous.

Alkali
A type of substance that feels soapy, for example chalk.

Battery
A battery contains chemicals that produce electricity when they react together.

Circuit
A complete path round which electricity can flow.

Concave
Curving inwards, like a cave.

Conduction (to conduct)
A way in which heat flows. When you hold a cup of tea, heat is conducted to your hand.

Contraction (to contract)
When something gets smaller.

Convection
A way in which heat flows through liquids and gases. A radiator heats a room by convection.

Convex
Curving outwards.

Current electricity
Electricity that moves or flows.

Energy
Something that makes things work. Heat, light and sound are all types of energy.

Expansion (to expand)
When something gets bigger.

Force
A push or a pull, for example when you kick a football.

Friction
A force created when two things rub together. It stops things or slows them down.

Gears
A simple machine with toothed wheels that slot together.

Gravity
A force that pulls objects towards the Earth.

Insulation (to insulate)
Materials that keep things warm or cold.

Lever
A simple machine that makes it easier to do things, for example a nutcracker.

Matter
Everything that exists in the universe. The three types of matter are liquids, solids and gases.

Mixture
Something made from several different pure substances.

Opaque
Something that light cannot pass through.

Oxide
A combination of oxygen and another substance such as iron.

Pendulum
A weight on the end of a rod or a piece of string that swings back and forth regularly.

Pressure
The force put on a surface. A force spread over a big area produces less pressure than when spread over a small area.

Pulley
A machine that makes it easier to lift heavy things.

Radiation (to radiate)
One of three ways in which heat flows. The Sun spreads heat (and light) by radiation.

Reflection (to reflect)
Something which is bounced off a surface. Sound and light can both be reflected.

Static electricity
Electricity that does not move.

Transparent
Something you can see through, for example glass.

Vapour
Another word for gas.

Vibration (to vibrate)
Something that shakes, or moves back and forth.

Volume
The size of something or the amount of space it takes up.

science
SCHOOL
Certificate
You have
completed
science
SCHOOL
with excellence,
Brita Granström – Tutor
Brita Granström